WILD FLOWERS
of FIELD *&* SLOPE
IN THE PACIFIC NORTHWEST

Written and photographed by
Lewis J. Clark

Edited by
John Trelawny

H A R B O U R   P U B L I S H I N G

Published by

HARBOUR PUBLISHING
P.O. Box 219
Madeira Park, BC Canada
V0N 2H0
www.harbourpublishing.com

Originally published: Sidney, B.C., Gray's Publishing, 1974
Cover photo: Pink rhododendron by Lewis Clark
Cover and interior design by Martin Nichols
Printed and bound in Hong Kong by Prolong Press Limited

Harbour Publishing acknowledges the financial support of the Government of Canada through the Book Publishing Industry Development Program (BPIDP) and the Canada Council for the Arts, and the Province of British Columbia through the British Columbia Arts Council, for its publishing activities.

THE CANADA COUNCIL | LE CONSEIL DES ARTS
FOR THE ARTS | DU CANADA
SINCE 1957 | DEPUIS 1957

National Library of Canada Cataloguing in Publication Data

Clark, Lewis J., 1907–1974.
   Wild flowers of field & slope in the Pacific Northwest

Includes index.
   ISBN 1-55017-255-7

    1. Wild flowers—Northwest, Pacific—Identification. 2. Meadow plants—Northwest, Pacific—Identification. I. Title.
QK144.C54 2002   582'.13'09795   C2001-911668-3

Introduction

In terms of geographic range, wild flowers included in this series are the more common or showy plants that occur from southern Alaska, through British Columbia, Washington, and Oregon, to northern California; from the coast to the western timber-line of the Rockies. A majority also occur in the Yukon, Alberta, Idaho, and Montana. Most of the flowers that inhabit fields and slopes will be found in this book, however some of those that are tolerant of a range of habitats have been included, because of space limitations, in the books of mountain, or arid flatland, or sea coast species.

Many of our earliest memories about flowers relate to the *places* in which different kinds grew. Who does not remember that in spring Buttercups dance at the lower end—the moister part—of a certain field. Or that, in the short turf of a woodland glade, delicious Wild Strawberries could be found!

Some plants (like some adaptable individuals) are happy in a variety of habitats. But others will only flourish within a narrow range of conditions—such as rainfall, drainage, shade, temperature, wind, and kind of soil. Do the kinds of flowers that grow in open, fertile fields, or on spring-wet slopes, display special adaptations?

It is a fair generalization that the more severe the environmental conditions, the more pronounced are the adaptations that plants must make to survive. Thus high alpine plants display very characteristic features, both external and internal, dictated by the rigorous climate of the mountains. Similarly plants subject to salt spray along the sea coast have evolved various structural and physiological responses. So do plants immersed—even for part of the year—in water. Perhaps the least spectacular adaptations are required of plants that inhabit the deeper, richer soil of fields and natural meadows, or of gentle slopes at low to moderate altitudes.

Yet on closer inspection we discover that even in such favourable environments, characteristic plants display some—rather generalized—adaptations. Precisely because conditions of soil and light in such habitats favour plant growth, there will be keen competition among plants of field and slope. The large available supply of nutrients in deep rich soil permits, in each growing season, vigorous and tall growth. But in such habitats wind is usual, and wind will limit plant height, or force such adaptations in herbaceous plants as a climbing or clambering habitat, e.g., using tendrils to take advantage of the woody stems of shrubs or trees. However, not all plants of open fields will be tall, or climbers.

The tall plants will create a whole series of micro-climates—at various levels—right down to the ground, and different species will respond to the minute variations at each level. An easily-observed response is period of flowering. Many plants occupying the 'ground-floor' (from the surface to 2 or 3 inches) come into bloom very early, before taller-growing species can cut off both light and readiness of approach by most of the pollinating insects. Thus Wild Strawberry will be in bloom while the neighbouring Columbian Lily is yet pushing its tall stem upward. The reverse strategy is employed by Harvest Brodiaea and Farewell-to-Spring, that wait until the lush spring growth of rock slopes has withered in the summer drought—before producing flowers.

Another interesting adaptation is shown by Western Long-spurred Violet, a low plant that produces early spring flowers. But in midsummer, when taller grasses and other over-shadowing plants claim the attention of bees, the Violet forms another type of (cleistogamous) flower, that never opens, and self-pollinates to form visible seed. Spring Gold opens its first flowers as early as December, on plants 1 or 2 inches high, but as adjacent herbage reaches for the sun, elevates its own stems to 12 or 16 inches, and continues to produce flowers until July. Wild Caraway uses its early leaves to store energy in its tuber-roots, then in August or even September, when the grasses are sere and brown, utilizes this stored energy to spread its white flowers from the top of a 2- or 3- foot stem.

About vacation-time, when children tumble out of the schools, millions upon millions of Ox-eye Daisies whiten the fields:

*'I saw the white daisies go down to the sea,
A host in the sunshine, a snow-drift in June.'*

—Bliss Carman: *Daisies*

We have all noticed that this plant dominates many fields, but few enquire the reason. Ox-eye Daisy extends creeping rootstocks in every direction, until the uniform-height stems that grow from them have suppressed competing species. A number of field and roadside plants have adopted the device of creeping rootstocks (rhizomes), such as St. John's Wort, Slender Speedwell, and the various Clovers.

Since winds can blow freely over open slopes and unobstructed fields, many plants that inhabit these areas produce, like the Fireweed, Goldenrod, and Oyster Plant, immense numbers of winged seeds, that parachute far and wide.

Some field species, like the Paintbrushes and the Bastard Toadflax, fight the severe competition by becoming partial parasites, and stealing part of their food from adjacent plants. Or, like the Broomrape, they may become total parasites, completely dependent upon the Stonecrops and other plants of rock slopes.

It is our observation that immigrant and invasive plants quickly take over broken or disturbed or waste ground, as along roadsides. Sorrels, Docks, Pigweed, White Campion, Bird's Foot Trefoil, St. John's Wort, Wild Carrot, Butter-and-Eggs, Tansy, Oyster Plant, Canada Thistle—invaders all, now dominate most of such disturbed sites.

This should bring to all of us a heightened awareness of the need for natural areas, parks, sanctuaries, environmental reserves. It is true that some of the invasive weeds are gradually replaced by native plants, but at best this is a slow process, and at worst—many of our choicest natives, once destroyed by logging, or road construction, or building developments, never return.

We have stressed common, or English, names, since there has been, traditionally, some resistance to exclusive use of the scientific names. There are considerable problems in determining what is *the* common name, for some species have a dozen or more local names, while others (as might be expected in a great expanse of recently explored country) have not yet received a common name. Clearly, we should have a generally accepted common name for every plant, just as (with few and decreasing exceptions) we have agreed upon one scientific name for each.

We are indebted—in a very large measure—for the taxonomy employed in this series, to Hitchcock, Cronquist, Ownbey, and Thompson, in their great 5-volume work *Vascular Plants of the Pacific Northwest*. For more northern species we are obligated to definitive studies by Calder and Taylor (*Flora of the Queen Charlotte Islands*) and by Hultén (*Flora of Alaska and Neighboring Territories*), also to the invaluable monographs by Szczawinski, and by T. M. C. Taylor, issued by the Provincial Museum, Victoria. Not included in this series are the trees or ferns (for which several excellent illustrated manuals are available) nor the horsetails, grasses, sedges, or rushes (which are of interest chiefly to specialists). However, an effort has been made to include in each book one or more plants that are representative of each of the major families of flowering plants.

The scientific names (given in parentheses) consist of a generic name, e.g. *Saxifraga* (meaning 'rock-breaker', since the roots pry apart crevices in broken rock) to which all the Saxifrages belong— plus a specific name, e.g. *integrifolia* (meaning with entire, or un-notched leaves), which identifies this one plant. (With very few exceptions, sub-specific and varietal names are not used in these guide books, unless precision demands them.)

In the brief descriptions we have used technical terms as little as possible: these few are illustrated in the Glossary at the end of the book.

Dates of flowering have not been given in all cases, because they vary widely over an area extending some 1500 miles from north to south, and from sea-level to about 8000 feet.

A plant identified becomes a known friend, to gladden the heart no matter how unfamiliar the path. And in a life-time one will constantly observe fresh details about each kind, and add the pleasure of discovery to enjoyment of their beauty.

When each of us has come to appreciate the inter-dependency of plants, and insects, and animals (including man), we will understand the necessity to preserve the fragile complex of the natural world.

x 2.0

1. **HOOKER'S ONION**, *Allium acuminatum*
This bright flower of the Lily family appears on short turf slopes that have
become sere and brown, and when its own 2-4 narrow leaves have shrivelled. It ranges west of the Cascades in B.C. and Wash., but is abundant
and general east of the Cascades from B.C. to Cal. The 3–10" flower-stems
support open clusters of as many as 25 flowers, which are occasionally
white. Bulb is ovoid, its outer scales marked in a regular checkerboard pat-
tern. WILD CHIVES, *A. schoenoprasum*, of damp slopes, Alas., e. B.C.,
Wash. to Cal., taller (8–20") with 2 persistently green leaves, has very slim,
variably purple flowers and a clump of slender bulbs. TALL SWAMP
ONION, *A. validum*, is very like the last, but has flat, not hollow leaves.
Cascades and Coast mts., s. B.C. to Cal.

x 0.75

x 2.0

2. **NODDING ONION**, *Allium cernuum*
The shepherd's crook at the top of the 4–16" stem at once distinguishes this common plant from numerous Onion species of our area. Widespread on rocky slopes, B.C., w. Wash., w. Ore. to Cal. SIERRA ONION, *A. campanulatum*, (5–15") is very beautiful, with open clusters of star-shaped flowers having pinkish-purple petals marked with a basal dark purple crescent. E. Ore. and Cal.

3. **LYALL'S BUTTERFLY TULIP**,
Calochortus lyallii
This beautiful species (4–20" tall) is very like **5**, but the petals are pointed-ovate, white, often flushed with purple, and conspicuously marked by a dark purplish crescent bearing fleshy hairs. Range extreme s. B.C., and Wash., e. of the Cascades, on dry and open slopes.

x 1.0

4. **HARVEST BRODIAEA**, *Brodiaea coronaria*
Grass-like leaves of this pretty 4–10" plant have withered by July or August, on sere hillsides from s. Vanc. Is., w. of the Cascades to s. Cal. They grow from rounded corms, once important as food for the Natives. Base of flower is plump and rounded, but slim and tapered in the slightly larger-flowered but very similar ELEGANT CLUSTER LILY, *B. elegans*, found from Ore. to Cal. HENDERSON'S BRODIAEA, *B. hendersonii*, is also from Ore. and Cal., and has slim yellowish flowers with flared lobes marked with purplish mid-veins. OOKOW, *B. pulchella*, whose bulbs were an important native food, grows 1–5' tall! Purple flowers in a compact ball-head. Open fields w. Wash. and Ore.

x 1.0

5. **THREE-SPOT TULIP**, *Calochortus apiculatus*
This beautiful (6–14") plant is fairly common on dry rocky slopes, sometimes in light shade, s.e. B.C., n.w. Mont., and n.e. Wash. The cream-
coloured cup-flowers are filled with long soft hairs, and distinctively
marked with a small round or purplish gland at the base of each petal. Also
distinctive is the shape of the anthers (upper part of stamen) which taper to
a slender point. SEGO LILY, *C. nuttallii*, of Mormon fame, is white, usually with reddish-purple markings above the yellowish base of each petal, but
is generally found on the Great Plains east of our range. SINGLE-FLOW-
ERED MARIPOSA, *C. uniflorus*, is shorter (4–8"), lilac-flowered, usually
with a small rounded purple gland, and nodding capsule. Fields, w. Ore.
and Cal.

x 2.0

6. OREGON CAT'S EAR,
Calochortus tolmiei
This white, cream, pinkish or
purplish species is abundant on
dry brushy slopes, Ore. w. of the
Cascades, to Cal. Plants are
3–8", usually branched, with 1–5
or more flowers whose inner
faces are entirely covered with
long fleshy hairs. Capsule is
nodding.

7. EARLY CAMAS,
Camassia quamash
This plant is familiar in fields
from B.C. to Cal., on both sides
of the Cascades. In some areas
Meriwether Lewis' description
(1806) is still applicable—'it
resembles lakes of fine clear
water.' Usually 5 of the flower
(perianth) segments curve
upwards and are more closely
spaced, the 6th downwards.
Buds are fewer and more sepa-
rated, and the plants less tall
and several weeks earlier blooming than **8**. The bulbs were
essential food for the early
Indian tribes.

x 0.3

x 0.3

8. **GREAT CAMAS**, *Camassia leichtlinii*

This fine plant occurs west of the Cascades, from s. B.C. to n. Cal., in meadows and moisture-holding crevices on rocky slopes. In good soil the stem reaches 3' or even 4', and bears 60–200 close-packed buds, which open to broad flowers having 6 regularly spaced sepals and petals. Characteristically, these twist together as they shrivel. White individuals are not uncommon, and the blue varies from pale to deep shades. Flowers of this and **7** are without scent. CUSICK'S CAMAS, *C. cusickii*, is restricted to n.e. Ore., in damp sub-alpine meadows. Its egg-shaped bulb may be 2" thick, and supports a very sturdy stem topped by very numerous and crowded powder-blue flowers, unpleasantly scented.

x 0.5

9. WHITE FAWN LILY, *Erythronium oregonum*
Occurs from Vancouver Island and extreme s.w. coastal B.C. to n.w.
California, west of the Cascades in fields, moist pockets in broken rock
ledges, and open woods (particularly under Garry oak). This fine species
lifts its huge white blooms 6–24" above 2 handsomely mottled basal leaves.
Usually solitary on its graceful stem, doubles and triples are common in
richer soil, and we have seen as many as 6 superb flowers on a stem.
Reverse side of sepals is variably yellow-green to rose-maroon, and the
choice orange and purple markings on the inner faces of the petals are usually more intense the better the exposure to light. Many of its former haunts
have yielded to the bulldozer, and if we wish our children's children to
enjoy this beautiful flower, we must provide sanctuaries for it.

x 0.7

x 0.75

10. CHOCOLATE LILY, *Fritillaria lanceolata*
Rising from a cluster of edible bulbs with many offsets (exactly resembling grains of white rice), a 6–36" stem bears slim pointed leaves arranged chiefly in 1 or 2 whorls. The 1–2 (rarely to 20) flowers nod, and are ill-scented. Range s. B.C. and Wash., but only w. of Cascades in Ore. and Cal. KAMCHATKA FRITILLARY, *F. camschatcensis*, (8–20") is very similar but has bell-shaped rather than the ball-shaped flowers of its more southern relative. Range w. Alas., BC, coastal Wash.

11. POISON CAMAS, *Zigadenus venenosus*
(formerly *Zygadenus venenosus*)
This 6–20" plant occurs with **7** and **8**, and its toxic bulbs not infrequently caused fatalities when mistaken by the Natives for the edible kinds, since they were dug before the distinctive flowers appeared. Flowers of this dangerous species are always cream-coloured, much smaller, and more crowded. Range B.C. to Cal.

x 0.2

12. COLUMBIAN LILY, *Lilium columbianum*
This attractive lily is still fairly abundant in damp woods and open meadows from central B.C., on both sides of the Cascades, south to n. Cal. The handsome orange flowers are ornamented with numerous maroon spots. Perianth segments (tepals) become strongly reflexed a few hours after the bud opens, and the blooms look modestly down from the top of a stem that may reach 4'. Whorls of 6–9 smooth lanceolate leaves occur along the stem.

x 1.6

13. **WILD HYACINTH**, *Triteleia grandiflora*

The lovely bluish-lavender tube-flowers are short, 1" long, arranged in an open cluster of 5–20 blooms. The smooth (10–24") stem grows from a fibrous-coated roundish corm, which also supports 1–2 narrow leaves. These are almost as tall as the flowering-stem (scape), and unlike those of **4**, remain green during the blooming period. Though for untold centuries it survived digging by Natives (who were essentially conservationists) the bulldozer and the plough have made this beautiful plant extinct over large stretches of its former range, from open Ponderosa woodland and range grassland, s. B.C. to s.e. Ore., chiefly east of the Cascades. Many will know this plant as *Brodiaea douglasii*. Our illustration is of the var. *howellii*, sometimes considered a separate species.

x 1.4

14. WHITE TRITELEIA, *Triteleia hyacinthina*
This common plant occurs in broken-rock outcrops near the coast to mid-
montane meadows, ranging from s. B.C. west of the Cascades to n. Cal.,
and also eastward from the latitude of Wenatchee. The rounded cluster of
papery-textured white flowers appear at the top of a smooth 10–28" stem,
from May to August. Backs of sepals and petals are marked by a vivid
greenish-blue line. Leaves usually remain green through flowering time.
BICOLOURED TRITELEIA, *T. howellii*, is very like **13**, but its blue tubular flowers are striped with deeper bluish-violet, with lobes that are white
and spread only slightly (never reflex). Occurs in coastal bluffs to sagebrush
slopes, Wash. and Ore.

x 0.5

x 0.5

15. OREGON IRIS, *Iris tenax*
This beautiful plant of prairies and open coniferous or oak forest, from s.w. Wash. to n.w. Cal., shows colour forms from white, cream, yellow, to blue and deep purple. Stems (10–14") bear 1–4 narrow leaves, and 2 purplish membranous leaves that shield the buds. Taller (14–24"), from much thicker roots (rhizomes), WESTERN IRIS, *I. missouriensis* bears pale to deep purplish-blue flowers on a stem that has 0–1 small leaves. It replaces the above east of the Cascades, and is spasmodic in B.C.

16. RED-FRUITED BASTARD TOAD-FLAX, *Comandra livida*
This 4–9" plant bears a few, inconspicuous purplish-green, flat, 5-sepal flowers. Small ($1/4$" or less), they lack petals, and yield to bright red fruit, containing a single large seed. Open woods, sub-alpine slopes, Alas. to n. Wash.

x 0.85

x 0.5

17. NARROW-LEAVED BLUE-EYED GRASS,
Sisyrinchium angustifolium
Extremely variable, this plant in some form is common from s. Alas. to s.
Cal., from coast to mid-mountain levels. It prefers places—wet in spring—in
fields and sagebrush deserts. Leaves are variously narrow, often shorter
than the flower-stems. Flowers are variably bluish to violet, usually about 1"
wide.

18. HELLEBORINE, *Epipactis helleborine*
This 6–20" immigrant plant is invasive and is becoming common in numerous areas in s. B.C. and n. Wash. Its crowded greenish-brown flowers have
an undivided lower lip, which is greenish, suffused with madder and rose.
GIANT HELLEBORINE, *E. gigantea*, may reach 4' and is distinguished by
the flower's lip, which is as long as the sepals, curiously constricted and
bent downwards with short erect lobes. Hot-springs and moist banks, B.C.
to s. Cal.

x 1.5

19. **SATIN-FLOWER**, *Olsynium douglasii*
(formerly *Sisyrinchium douglasii*)
This beautiful plant is unmistakable, for it blooms in early February, and its
gorgeous reddish-purple flowers are as much as 1¹/₂" across. (Rarely white,
pink, or striped individuals are found.) Flowers are borne on delicate stalks
(pedicels) which arise between a pair of bracts that are unequal in length.
The silvery-green, dark-veined leaves grow in tufts, and are usually shorter
than the flower-stem, which is 6–16" tall. The delicate bells nod from rocky
ledges near the sea to sagebrush slopes and up to 6000' elevation. Range s.
B.C. to Cal., and—in a form often paler, with a more inflated base of the
anther-tube—also on the e. side of the Cascades.

x 0.4

20. YELLOW LADY'S SLIPPER, *Cypripedium calceolus*
The lovely flowers of this magnificent orchid may be found from April to
July, sometimes in crowded clumps, from n. B.C., through Wash. and Ore.
(east of the Cascades). Over its enormous circumboreal range it is quite
variable in size, scent, and colour of the side petals, which may be yellowish, greenish to purplish-brown, and are variably twisted. The inflated
'pouch', in all forms, is golden-yellow. Commonly a single elegant and perfumed bloom is held 6–12" above the broad and conspicuously-veined
leaves. Prefers damp, open slopes, bogs, and gravelly outwashes, and is
absent from the coast.

x 0.75

x 0.85

21. HOODED LADIES' TRESSES,
Spiranthes romanzoffiana
The small, greenish-white, faintly perfumed flowers are closely packed in 3 spiral ranks, like neatly braided hair. The parallel-veined leaves are chiefly basal, though a few small ones appear on the 4–16" unbranched stem. Range is extensive, from Alas. south to Cal., and east to Nfld., in moist places in fields and swamps.

22. COMMON SORREL,
Rumex acetosella
The long yellow rootstocks of this pernicious weed are only too familiar in our gardens. Smooth, sour-tasting leaves are shaped like mediaeval halberds. The minute reddish flowers are either staminate or pistillate. Like many sorrels and docks, this is an immigrant from Europe. GARDEN SORREL, *R. acetosa* is much taller (to 3') with larger arrow-head shaped leaves. Sporadic in distribution.

x 1.5

x 0.5

23. BASTARD TOAD-FLAX, *Comandra umbellata*
The 3–12", clustered, leafy stems of this plant parasitize the roots of a variety of plants from dry coastal ridges to sagebrush desert over most of N. Amer. Western plants usually have thick greyish leaves. Flowers are white, cream, or purplish, carried in clusters, and yielding to bluish- or purplish-brown berry-like fruit.

24. PIGWEED, *Amaranthus retroflexus*
Coarse (20–36") and stiff-haired, this weed is common on waste ground and roadsides all over the continent. The tiny flowers are densely crowded, and bear 0–5 stamens or pistils, or both. WHITE PIGWEED, *A. albus*, (whitish stems) is also widespread and weedy, but its equally insignificant flowers occur in much smaller clusters *in the axils of the leaves*. Range B.C. to tropical America.

26

x 0.4

25

x 0.5

25. **WESTERN DOCK**,
Rumex occidentalis
This is often a striking plant, since leaves, stems, and especially the fruit, turn rosy-red as the summer advances. The fruit has broad valves that are conspicuously net-veined. Range Alas. to c. Cal., in marshes, and summer-drying fields. Probably 14 other Docks occur in this large area. They are largely distinguished by microscopic details of the fruit.

26. **MOUSE-EAR CHICKWEED**,
Cerastium arvense
This far-ranging and plastic (2–18") species is common, from rocky slopes by the coast to sandy river bars, dry inland fields and even sub-alpine meadows all over the continent, n. of Mex. Leaves and stems are usually furred with white hairs. Flowers may be nearly 3/4" across, with petals lobed one-third of their length. Pesky GARDEN CHICKWEED, *Stellaria media*, has a 3-cleft pistil (instead of 4–5 cleft).

x 0.85

x 0.5

27. WHITE CAMPION, *Lychnis alba*

Weedy but cheery along roadsides and in waste ground, this ubiquitous immigrant bears staminate and pistillate flowers (having a 5-cleft pistil) on separate plants. Pointed-elliptical, hairy leaves are most numerous near the base of the branching, 20–48" stems. Similar weedy BLADDER CAMPI-ON, *Silene cucubalus*, has a 3-cleft pistil, and smooth, veined 'bladder' calyx-cups.

28. MULLEIN PINK, *Lychnis coronaria*

This immigrant and showy species has escaped widely from gardens, along roadsides, chiefly w. of the Cascades, B.C. to Ore. The claret-red flowers are striking above the silver-hoary foliage. Similarly coloured is CORN COCK-LE, *Agrostemma githago*, distinguished by broadly oval leaves and 2 point-ed ears midway on each petal.

x 0.75

x 1.5

29. HOOKER'S PINK, *Silene hookeri*
This very beautiful, low (2–6"), sprawl-ing plant has greyish, 23", lanceolate leaves that make a perfect foil for the lovely pale- to deep-pink flowers. Found on open hillsides, and edges of fields, but never commonly, from the e. side of the Coast range, Ore. to n.w. Cal. SMALL-FLOWERED CATCHFLY, *S. gallica*, (4–16") is extremely sticky with white to palest-pink flowers, 1/4" or less. B.C. to Cal., w. of Cascades.

30. WESTERN BUTTERCUP,
Ranunculus occidentalis
This is probably the most abundant of many buttercups that occur in fields and rock-slopes, Alas. to Cal. A myriad of its bright cups gild fields of Camas in mid-May. Its greenish sepals are each spread flat except for the outer half, which is sharply reflexed; the 5–8 petals have an egg-shaped nectary scale at each base.

x 1.6

31. **MONKSHOOD**, *Aconitum columbianum*
This interesting plant bears curiously-shaped, very dark flowers in an open
spike, at the top of 2–7' erect stems. These stems are hollow and usually
smooth, with large, long-stalked, maple-shaped leaves alternating from the
ground upward. Much reduced ones even extend into the lower part of the
flower-spike. Found in moist woods and sub-alpine meadows, Alas. to Cal.,
coastal mts. to Rockies. The whole plant is poisonous, especially the seeds
and the swollen roots. Only 6–24" tall, NORTHERN MONKSHOOD, *A.
delphinifolium*, is 1- to few-flowered, and at least as poisonous. It occurs
in Alaska, the Queen Charlotte Islands, and the northern Rockies.

x 0.2

32. **WESTERN COLUMBINE**, *Aquilegia formosa*
This lovely plant is widely distributed from Alas. to s. Cal., from the sea-coast to moderate heights in the central mts. It flourishes in gravel bars, moist crevices among the rocks, and in light shade on the fringes of the forest. Beautifully lobed leaves almost suggest those of maidenhair ferns. The 5 coral-red sepals enhance the bright yellow of elegantly flared tips of the 5 curious petals. These petals become coral and tubular above, finally swelling to nectar-filled enlargements. Hummingbirds and long-tongued swallowtail butterflies come to sample this hidden nectar.

x 1.0

33. MENZIES' LARKSPUR, *Delphinium menziesii*
This is the abundant species on the west side of the Cascades, B.C. to Cal.
Generally about a foot high, in impoverished or highly favourable circum-
stances it can be 4" or 30". Colour is variable from (rarely) white to shades
and tints of blue to violet. The 2 upper fan-shaped petals (blue in the illustration) are often white. Characteristic are tuberous-clustered roots, shal-
lowly-notched petals, and lower flower-stalks (pedicels) much longer than
the flowers themselves. Provides bright accents on broken rock slopes and
in open oak woodland. TWO-COLOURED LARKSPUR, *D. bicolor*, is a
small species with thick basal leaves deeply cut into overlapping segments;
the 2 upper petals are small and pale-blue, pencilled purple. Olympics, oth-
erwise e. of Cascades.

x 0.75

34. **TALL OREGON GRAPE**, *Mahonia aquifolium*
(formerly *Berberis aquifolium*)
A treasure plant in many ways, this superb shrub is widespread on rocky
slopes, open woods, and even sagebrush slopes, from s. B.C. to c. Ore.,
westward from the e. base of the Cascades. This fine species is outstanding,
not only for size (1–10'), handsome leaves, and edible blue fruit, but also
for its quantities of beautiful yellow, honey-scented flowers. Inner bark is
bright yellow and yields a fine dye. On close examination the flowers
remind one of tiny short-trumpet daffodils. CREEPING OREGON
GRAPE, *M. repens* (formerly *Berberis repens*), is very similar but sprawl-
ing, and usually about 1' high. Leaves are more bluish, and the abundant
fruit appears much darker blue-black, because the pale dusty 'bloom' tends
to be sparse. From B.C. to Cal. it replaces **34** east of the Cascades.

x 0.5

x 1.5

35. **CHARLOCK**, *Sinapsis avensis* (formerly *Brassica kaber*)
This weedy immigrant (12–36" tall) is found in waste places and fields
throughout our area. The pointed, smooth, upright seed pods (siliques) are
tipped with a flattened beak. Even more common is FIELD MUSTARD,
Brassica rapa (formerly *Brassica campestris*), distinguished by silique
beaks that are round in section, and smooth leaves that clasp the stem. It
yellows neglected grain fields from Alas. to Cal., eastward.

36. **TUFTED SAXIFRAGE**, *Saxifraga caespitosa*
The tuft of distinctive, finely glandular foliage identifies this 3–5" plant,
whose distribution is world wide, including Alas. to s. Ore. The leaves are
bundled in little whorls and are wedge-shaped, but with 3 finger divisions
at the tip. Highly variable, the white flowers have 5 round to notched
petals, but the anthers are always longer than the sepals. This compact
plant inhabits rock slopes from sea level to very high altitudes.

x 0.6

x 0.75

37. **DAME'S VIOLET**, *Hesperis matronalis*
This tall (20–48"), rather coarse plant frequently escapes from gardens,
almost anywhere in our area, on roadsides and waste ground. Flowers vary
from white to pink to purple. Their scent, especially at dusk, is pleasant.

38. **EARLY SAXIFRAGE**, *Saxifraga integrifolia*
Very different to **36**, is this rather coarse-leaved and small-flowered species
of rock benches and grassy slopes to sub-alpine meadows from B.C. to Cal.
Conspicuous are the red gland-tipped hairs of the flower-stalk, which is
4–12" tall. In flower as early as February.

x 1.0

39. **BROAD-LEAVED STONECROP**, *Sedum spathulifolium*
This is a plant of many virtues—evergreen, hardy, able to creep up vertical
rock-faces yet never invasive, spreading a low canopy of dazzling golden
flowers over beautiful foliage-rosettes. The thumb-shaped, fleshy leaves
show a dozen hues and often turn a brilliant rose colour during the winter.
The species is easily recognized: sepals are blunt-tipped and joined at their
extreme bases; the mature seedcases are slim-pointed, upright but with
extreme tips sharply flared outward. Found chiefly on rocky ledges from sea
coast to low altitudes in the Coast and Cascades ranges, from s. B.C. to Cal.

x 0.5

40. **SMALL-FLOWERED ALUMROOT**, *Heuchera micrantha*

Airy clusters of small, delicate white flowers lifted up 10–24" above notched, heart-shaped basal leaves characterize this attractive member of the Saxifrage clan. The leaves are usually slightly longer than broad and somewhat hairy, with long stalks (petioles) nearly always conspicuously white- or rusty-haired. The hairy calyx-cup is one-third notched into 5 lobes. There are 5 stamens. Grows from cliff faces, on rock slopes, and gravel bars, B.C. to Cal., chiefly west of the Cascades. SMOOTH ALUMROOT, *H. glabra*, prefers slightly wetter slopes, Alas. to Ore., coast to Rockies. Its leaves are usually broader than long, and smooth, like the petioles.

x 1.2

41. **RED-FLOWERED GOOSEBERRY**, *Ribes lobbii*
This very attractive spiny 2–4' shrub in early summer bears showy flowers,
in fall quantities of large, reddish-brown gooseberries. Though occurring
over a wide range B.C. to Cal., w. of the Cascades (except in Wash. where
on both sides), the plant is, apparently, nowhere abundant. It prefers open
rocky slopes from sea level to moderate altitudes. The fuchsia-like flowers
are unmistakable; the $3/4$", egg-shaped, unpalatable fruit is covered with
short glandular hairs and tipped by the reddish-brown, withered residue of
the flower. COAST BLACK GOOSEBERRY, *R. divaricatum*, (B.C. to Cal.,
coast to Cascades) has similar but much smaller, variably whitish to reddish
flowers, triple pale brown thorns, and $1/2$" smooth, black, nearly round,
palatable fruit.

x 0.85

42. FLOWERING RED CURRANT, *Ribes sanguineum*

This is one of our most beautiful shrubs. From 3–9' tall, it is spectacular in April–June when literally covered by pendulous clusters of lovely flowers. Though the specific name means blood red, the colour of the 5-lobed calyx-tube varies from bluish-pink to an intense and fiery carmine. (Very rarely a superb pearl-white form is seen). The petals are small, erect, and usually white, rarely pink. The unpalatable fruit is nearly round, nearly black but covered by a pale blue 'bloom.' Characteristic are the reddish-brown bark, and irregularly 5-lobed leaves, crinkled as if cut from crepe paper. Leaves are paler and more hairy beneath. Widespread w. of the Cascades from B.C. to Cal.

x 0.8

43. MOCK ORANGE,
Philadelphus lewisii
Along the coast, specimens
10–12' tall are beautiful accents
of coastal thickets, as well as
rocky hillsides in sagebrush
deserts. In June they bear a pro-
fusion of perfumed showy blos-
soms. The deciduous leaves are
about 3" long, with 3 main veins
and a few edge serrations.
Usually 4 oblong petals enclose
about 30 uneven stamens. B.C.
to Cal., coast to Rockies.

44. BITTER CHERRY, *Prunus
emarginata*
Often a straggly shrub in the Dry
Interior, in coastal woods this is
a 30–50' tree. Its papery bark,
marked by lenticels, is very cher-
ry-like. Leaves are about 2" long,
finely saw-edged. Fruit is very
bitter, red, less often black. Bark
of young twigs is reddish-purple.
Range is B.C. to Cal., coast to
Rockies.

44

x 0.7

45

46

x 1.2

x 0.6

45. WILD STRAWBERRY, *Fragaria virginiana*
At times liberally sprinkling short turf from Alas. to Cal., east to the
Atlantic, the delicious fruit of this low plant is known to every country
child. The seeds are sunken 3/4 of their depth into the rounded red fruit,
and the leaves are thinner than those of COAST STRAWBERRY, *F.
chiloensis*, which has leathery leaves, and whose seeds scarcely dimple the
surface of the fruit.

46. GRACEFUL CINQUEFOIL, *Potentilla gracilis*
This extremely variable, and very common plant ranges from Alas. to Cal.,
coast to Rockies, in coastal flats to alpine meadows. From a cluster of
palmate, long-petioled, basal leaves rise 16–32" branched stems, having
usually 1–2 stalkless leaves. Stamens commonly number 20, and the slender
pistils are roughened only near the base. Belongs to Rose, not Buttercup
family.

x 0.6

47. PACIFIC CRABAPPLE, *Malus fusca* (formerly *Pyrus fusca*)
This attractive shrub (10–36' tall) forms dense thickets along stream banks
and swamp edges, which may be almost impenetrable because the interlac-
ing branches carry numerous spine-like spurs that are 1–2" long. The
extremely varied leaves turn brilliant shades of red and yellow in the fall,
when showy yellow fruit, sometimes red-cheeked, makes each bush spectacular. The calyx-cup has 5 pointed lobes, that are reflexed. Stamens are
about 20, pistils 3–4. Range is from coast to Cascades, s. Alas. to n. Cal.

x 0.75

x 1.3

48. **CLUSTERED WILD ROSE**,
Rosa pisocarpa
Characteristically, the buds and open
flowers occur in clusters of 3–6. This
beautiful shrub (3–7' tall) has prickles
that are slender, straight, and few. Leaf-
edges have very small notches. The
sepals are persistent and covered with
numerous, fleshy, hair-like protuber-
ances, each swollen-tipped. Flowers, a
beautiful soft pink, are 1^1/2"–2". Range
w. of Cascades, s. B.C. to n. Cal.

49. **RED CLOVER**, *Trifolium pratense*
This very familiar forage plant of cultivated fields and roadsides, produces
sweet honey, a fact known to every
child who has sucked the red florets.
Leaves frequently have whitish crescent mark, and are usually downy with
fine white hairs. Occurs in inhabited
areas from Alas. to Cal.

x 0.7

x 1.2

50. BIRD'S FOOT TREFOIL, *Lotus corniculatus*
This European pasture plant is now widely dispersed along roadsides throughout our area. Usually prostrate, to about 12", the sprawling stems bear numerous unstalked, hairless, trifoliate leaves, with a lower pair of smaller oval leaflets. Flowers 3–12 in a head, bright yellow often tinged with red. Long pods are arranged like a bird's foot.

51. BICOLOURED LUPINE, *Lupinus bicolor*
This little annual is abundant on well-drained soil near the coast, from Vancouver Island, and the Fraser River, southward to Cal. Its hairy rosettes appear soon after the new year and grow 2–12" tall. The central band on the blue banner petal is white, later turning pinkish, and marked by black dots. The keel is hairy on the upper half, and its slender tip curves upward.

x 0.3

52. LARGE-LEAVED LUPINE, *Lupinus polyphyllus*
The stately spires of this perennial lupine beautify wet roadside ditches, and moist ground to lower levels in the mountains. The flowers are very numerous on a crowded spike that may be 18" long, and variable in shades of blue to violet. Stems are tallest (to 4') of our lupines, and hollow. The leaflets in each long-stalked, palmate leaf number 10–17, and are notably broad ($1/2$–1"). Only the extreme tip of the keel, which is nearly always smooth, projects beyond the clasped wing-petals. Pods are 1–2" long, and densely covered with long hairs. Range from Alas. to Cal., from coast to Rockies.

53

54

x 0.7

x 1.0

53. HARESFOOT CLOVER, *Trifolium arvense*

This softly-haired, 6–12" annual was once grown as a fodder-plant, but has been displaced by higher-yield species. It may be found by roadsides and in waste ground throughout our area, though chiefly w. of the Cascades. Leaves are narrow and bear minute teeth. Calyx-lobes are thin and soft, white to fawn-coloured, extremely hairy, and much longer than the pale pink corolla.

54. WHITE CLOVER, *Trifolium repens*

This nearly smooth, immigrant plant is now found in waste ground throughout our area. The white flowers often age pinkish, and have smooth, sharply-pointed calyx-lobes. The leaves have long stalks and 3 fine-toothed leaflets. COMMON YELLOW TREFOIL, *T. dubium*, is a creeping annual, infesting lawns everywhere. Middle leaflet has a trifle longer stalk than the other 2, and often turns purplish.

x 1.0

HABITAT

55. LANCE CLOVER, *Tridentatum willdenowii*
(formerly *Trifolium tridentatum*)
This native annual species is abundant in pockets of soil on rocky hillsides,
and in short-turf meadows, w. of the Cascades, from B.C. to Cal. Glabrous
(smooth) throughout, the 36" plants bear few to many flowers, that are
variably purple (but tipped with white, or sometimes darker purple). Calyx-
lobes are sharply triple-pointed. Leaflets very narrow, and sharply-toothed.

56. DEER BRUSH, *Ceanothus integerrimus*
Tall (3–12') and freely-branched, this pretty shrub bears masses of tiny flowers that vary from pale blue to white. The 1–2", unnotched leaves are deciduous, and usually have 3 more prominent veins. Range e. side of Cascades,
Wash. to Cal. MAHALA MAT, *C. prostratus*, is a superb, low, matted plant
with holly-like, heavily-notched, 1/2–1" leaves, and small, flat clusters of
greyish-blue to white flowers. Wash. to Cal. chiefly e. of Cascades.

x 0.75

x 4.0

57. COW VETCH, *Vicia cracca*
This slightly downy perennial festoons roadside bushes with 2–4' stems
bearing 2–4" long leaves. Each leaf has 12–22 narrow leaflets and a long
branched tendril. The showy one-sided spikes of bright flowers have long
stalks. GIANT VETCH, *V. gigantea*, climbs to 12', and has hollow stems.
Leaves have 19–29 narrow leaflets, and inconspicuous greenish- to orange-
yellow flowers, sometimes purple-flushed. Alas. to Cal., esp. near coast.

58. DOVE'S-FOOT CRANE'S-BILL, *Geranium molle*
Early in the spring, the attractive rosettes of roundish (but deeply-lobed)
leaves are common on moist waste ground throughout our area, esp. w. of
the Cascades. Sepals are soft-haired, pointed but not bristle-tipped.
SMALL-FLOWERED CRANE'S-BILL, *G. pusillum*, of similar range, has
smaller, more bluish-lilac flowers having only 5 (of its 10) stamens that
develop fertile anthers (cf. 10 fertile stamens in **58**).

59

x 0.5

60

x 0.4

59. BUCKBRUSH, *Ceanothus sanguineus*
The specific name calls attention to the reddened flower-stalks, which contrast strikingly with the clusters of cream-white flowers. This open-branched deciduous shrub (3–9') has smooth, purplish branches bearing alternate leaves that have distinct stalks. Leaves are oval, glandular, and saw-toothed, but not as thick-textured as those of SNOWBRUSH, *C. velutinus*. The sticky varnished leaves of the latter suggest that it inhabits dry areas east of the Cascades. Both B.C. to Cal.

60. CASCARA, *Rhamnus purshiana*
Grouped, erect, silver-grey barked stems reach 15–30'. The bark was formerly much sought for the pharmaceutical trade. Leaves are 3–4" and unmistakable. Occasional black berries succeed the inconspicuous flowers. Range B.C. to Cal., chiefly w. of the Cascades.

61 | 62

x 1.0 | x 0.9

61. ST. JOHN'S WORT, *Hypericum perforatum*
This troublesome but handsome invader now lines countless miles of roads, and extensive areas of waste ground, from B.C. to Cal. Erect but much-branched stems (12–36") have numerous, opposite, entire-margined 1" leaves, and flattish clusters of yellow flowers. Margins of the 5 petals are peppered with tiny black dots. Sepals are narrowly-lanceolate.

62. FAREWELL-TO-SPRING, *Clarkia amoena*
This beautiful annual spreads its painted cups when spring is over. Wiry stems (3–30"), with 1–2¹/2" lanceolate leaves, support flowers that vary a great deal in size (up to 2¹/2"), colour (cream to rose), and markings (lined or suffused, or boldly splotched with purple-carmine). Found on summer-dry rock slopes w. of the Cascades, from S. Vanc. Is. to n. Cal.

x 1.5

63. WESTERN LONG-SPURRED VIOLET, *Viola adunca*
Meadows, open woodland, and moist roadsides are hosts to this very wide-spread purple Violet, that ranges over most of North America. Over this immense area the species is highly variable. Leaves are smooth, or hairy, usually heart-shaped. Flowers are blue to deep violet, the 3 lower petals whitish at base, penciled with purple lines. The 2 lateral petals are white-bearded near the inner end; the pistil (style) is also bearded near the tip. The spur is generally rather slender, and more than half as long as the lowest petal. Very similar is HOWELL'S VIOLET, *V. howellii*, limited to the w. side of the Cascades, s. B.C. to n. Cal., and distinguished by paler flowers with a much broader spur, that is much less than half as long as the lowest petal.

x 0.1

64. **FIREWEED**, *Epilobium angustifolium*
Who does not know this fine plant that speedily covers the blackened
residue of slash-fires with its dense stands of vivid rose-pink flowers? The
smooth stems (containing a sweet edible pith) are 4–8' tall, clothed to the
top with alternate, narrowly-lanceolate, smooth leaves. A conspicuous fea-
ture of the showy flower is the very long, drooping, white pistil. During
Indian summer days the air, over vast expanses of logged-off land, is filled
with the drifting white seed-parachutes. This beautiful species is widespread, from Alas. to Cal., eastward across the continent.

x 0.4

65. WATER HEMLOCK,
Cicuta douglasii
The tiny greenish-white flowers, in a somewhat flattened cluster, top 2–7' smooth, hollow stems (that are usually purplish streaked). Very large pinnately-compound leaves and globular ridged fruits help to identify this plant, whose swollen roots are dangerously poisonous. Prefers wet stream banks and meadows, Alas. to Cal.

x 0.5

66. POISON HEMLOCK,
Conium maculatum
This most unfortunate immigrant species is very poisonous, and spreading rapidly along roadside ditches and moist waste ground, over most of N. Amer. The smooth, hollow, 6–10' stems are 'maculate' (marked with elongated purple spots). The individual white florets have stalks that are notably uneven in length. Flowers have a faint, unpleasant, musky odour.

x 0.5

x 0.4

67. QUEEN ANNE'S LACE, *Daucus carota*

This wild carrot is a common weed of neglected fields and roadsides all over North America. Despite its invasiveness the flat-topped white flower-clusters and 'carrot' foliage are quite appealing. Usually there is a single purple flower in the centre of the cluster. As seeds mature the cluster infolds to form a perfect 'bird's nest.'

68. WILD CARAWAY, *Perideridia gairdneri*

The tall (16–40") solitary stems of this interesting plant bear extremely 'skinny' pinnate leaves. The lower ones have shrivelled before the sweet-smelling flower-clusters open. The swollen root (called Yampah) was a favourite Native food. The plant is in flower in Aug.–Sept., on open slopes and seasonally dry flats from s. B.C. to s. Cal., coast to considerable elevations in the Rockies.

x 0.5

69. **COW PARSNIP**, *Heracleum lanatum*
Heracleum is well named, for these are indeed plants of herculean propor-
tions, with giant leaves and hollow stalks from 4–12' in height. Stems and
compound leaves—each with 3 maple leaf shaped leaflets, as much as 12"
broad and long—are usually covered with white hairs (very sparsely so in
some forms). The flattened flower-cluster may be 8–10" across; it is made
up of small flowers usually greenish-white, occasionally yellowish, or even
pinkish. Outermost flowers in each compact sub-cluster have 4 deeply
notched petals that are very uneven in size. The heavily ribbed, flattened
fruits have a rather pleasant aromatic odour. The plants are browsed by cat-
tle, as well as by elk and deer. Widely distributed in moist low ground, coast
to moderate elevations in the mountains, Alas. to Cal. and eastward.

x 0.4

x 1.0

70. SPRING GOLD, *Lomatium utriculatum*
This much-loved, spritely member of the Umbellifer family is in bloom from
Dec. to mid-July! On the mossy covering of open rock slopes, Spring Gold
unfolds its attractive, soft, much-dissected, fern-like leaves—which are early
crowned with a level-topped cluster of bright yellow flowers. The flattened
fruit is smooth and elliptical, broadly winged on the edges, somewhat
ribbed on the faces. Range w. of Cascades, s. B.C. to Cal.

71. SIERRA SNAKE ROOT, *Sanicula graveolens*
Widespread from s. B.C. to s. Cal., and from sea level to moderate elevations in the mountains, this low (2–14") perennial lifts rounded yellow
flower-heads above deeply-lobed leaves. The small fruits are nearly round in
cross-section and densely covered with tiny hooked prickles that readily
cling to cuffs and stockings. There are more staminate than pistillate (and
perfect) flowers in each cluster.

x 0.6

x 0.6

72. MANZANITA, *Arctostaphylos columbiana*
The polished, smooth and rich mahogany-brown branches of this very handsome 3–10' shrub are so distinctive that there is no question of its identity. Thick, firm, elliptical leaves are grayish-green and set off handsomely the pink flowers, which are succeeded by shiny coffee-coloured berries. Range s. B.C. to c. Cal., coast to Cascades.

73. FEW-FLOWERED SHOOTING STAR, *Dodecatheon pulchellum*
(Earlier name, *Dodecatheon pauciflorum*)
Pulchellum means 'very beautiful'—an apt description. This species holds its slimmer leaves (see **77**) almost erect in a ring around the flower-stalk, and the tube of joined stamens is always yellow. Petals vary from pale to intense rose-pink. Seeds escape through a ring of slits at the upper end of the polished, pale brown seed-capsule. Range Alas. to Mex., coast to alpine slopes.

x 1.5

x 0.4

74. **BEARBERRY**, *Arctostaphylos uva-ursi*
This evergreen mat-like shrub is both ubiquitous and highly ornamental. It is abundant on rocky outcrops, as well as the dry floor of open woods from coastal Cal. n. to Alaska, and eastward. Pretty pink bells precede the handsome scarlet berries. The leathery leaves were dried and used to extend the tobacco of early explorers (who called it Kinnikinnick).

75. **HOUND'S TONGUE**, *Cynoglossum grande*
The coarse hairy leaves (up to 8" long, on equally long stalks) may suggest the common name. Smaller ones, finally lacking stalks (petioles), grow up the lower half of the 2–3' stems—but all are harsh-hairy on their lower surfaces. The 1/2" blue flowers are marked by a distinctive white, raised ring enclosing the 5 short stamens and central pistil. Range s. B.C, to Cal., w. of Cascades.

x 0.5

76. **PINK RHODODENDRON**, *Rhododendron macrophyllum*
This splendid 4–12' shrub is the state flower of Washington. It is at its best
in open thickets, chiefly coastal and never beyond the Cascades, s. B.C. to
n. Cal., where it forms a compact rounded bush 4–6' high, covered with
intensely coloured flower-trusses. This height is doubled where light is limited, as around deer-pools in forests—and in such tall specimens flowers are
sparse and pale. The leathery, elliptic-oblong leaves are about 5" long, evergreen and glossy-shining above. There is much variation in the beautiful
flowers, between pink and deep rose, with deeper or lighter spotting in the
throat. This shrub is protected by law in B.C. and in Wash.

x 0.6

77. BROAD-LEAVED SHOOTING STAR, *Dodecatheon hendersonii*
This beautiful species (also known as Peacock) is limited to the coast-to-
Cascades strip, southward from s. Vancouver Island to s. Cal. It is readily
distinguished from **73** by its broader leaves (broadest at a point nearer the
tip), which are held flat on the ground. The cyclamen-shaped flowers are
variably coloured, but usually more rose-purple. The tube formed from the
joined bases of the stamens is purple. The seed-capsules open quite differ-
ently in the 2 species. WHITE SHOOTING STAR, *D. dentatum*, has
strongly notched leaves and white petals with yellow anther-tube. It prefers
wet seepages to shaded moist slopes, east of the Cascades, in s. B.C., Wash.
and n.w. Ore.

78

79

x 0.85

x 0.2

78. DOGBANE, *Apocynum androsaemifolium*
A branching 12–24" plant with opposite-paired egg-shaped leaves (that droop characteristically in the heat of mid-day), this is common over most of U.S. and Canada, on dry, hard soil to moderate elevations. Dry pendulous fruiting-capsules may be 3" long. INDIAN HEMP, *A. cannabinum*, also exudes a milky juice from broken stems, but grows to 5', and has slim sharp-pointed leaves that are always erect. Similar range.

79. COMMON RED PAINTBRUSH, *Castilleja miniata*
This plant's splashes of vivid colour never fail to evoke a primitive excitement, whether in alpine meadow or stony lakeshore. It is variable in colour and in hairiness, 8–30" tall, with slim-lanceolate leaves mostly entire, though the upper ones are 2–3 lobed. There are many species of Paintbrushes, which are very difficult to distinguish. Most are partially parasitic. Western N. Amer., often subalpine.

x 1.0

80. **LARGE-FLOWERED COLLOMIA**, *Collomia grandiflora*
This beautiful annual plant is at once noticeable for the unusual colour of
its ball-head of flowers—a soft salmon-orange, occasionally more yellowish.
The 1–3' plants are usually unbranched, with numerous, smooth, unstalked,
lanceolate and entire leaves that alternate up the stem. If one of the flared
corolla-tubes is split one sees that the 5 stamens spring from the inner wall
of the tube at different levels, and that the stamens are remarkably uneven
in length. The attractive flowers may be seen from June to August, in open
woodland or dryish open slopes, from sea level to considerable altitudes.
Range s. B.C. to Cal., both sides of the Cascades. NARROW-LEAVED
COLLOMIA, *C. linearis* is very common over the same range. It is evil-
smelling, about 1/2 the size of **80**, with small flowers pink to bluish to
white. Leaves are sticky-haired.

x 2.0

81. **SPREADING PHLOX**, *Phlox diffusa*

A very showy sprawling mat-plant that splashes its variously coloured masses of bloom over dry rocky slopes from moderate to high altitudes in the hills and mountains, from s. Vancouver Island to n. Cal., east to the Rockies. Within a few square yards one may find colour forms from white to clear pink to various shades of lavender, purple, violet, and magenta, some with the petals marked basally with white, or yellow, or with a deeper shade of the outer petal colour. The flat lobes of the corolla also vary in width, from wheel-like, broad and overlapping—to narrow and spoke-like. Leaves are smooth, linear but not harsh, with tiny woolly hairs at the base of each.

x 0.5

x 2.5

82. **BITTERSWEET**, *Solanum dulcamara*
This immigrant vine-like plant may clamber for several yards through
shrubbery, or sprawl along the ground. It is occasional through our range,
preferring moist ground. It is a perennial, with large clusters of bluish-violet
flowers and bright scarlet fruit. BLACK NIGHTSHADE, *S. nigrum*, is a
bushy annual that in some of many variations occurs in moist soil around
the world. Leaves are pointed-oval, irregularly blunt-toothed, often hairless,
and frequently purple-edged. In this species the corolla is smaller and
white-lobed, followed by ultimately black berries.

83. **THYME-LEAVED SPEEDWELL**, *Veronica serpyllifolia*
This is a low and compact perennial widely distributed in short turf (and as
a garden weed) with oval, shiny, untoothed (or only slightly toothed)
leaves. Stems are very short-haired, but the leaves are almost hairless.
Flowers are terminal, in one variety bright blue (illustrated), in another pale
blue to white, marked by darker lines.

84. GOLDEN PAINTBRUSH, *Castilleja levisecta*
This very bright yellow Paintbrush may be found in fields and prairies at
low altitudes west of the Cascades, from Vancouver Island to Oregon. Like
others of its numerous relatives it is a perennial, and parasitic, at least to
some extent, upon the roots of other plants. This species has several 4–18″
stems, erect above but often at their bases inclined partly along the ground.
The whole plant is covered with soft white hairs, and the leaves are pointed
lanceolate and entire-edged below, but above both the leaves and the yellow bracts enclosing the flowers end in 2–4 pairs of short lobes. These yellow bracts need to be pulled back to reveal the flowers, which are very slim,
narrowly tubular, with a nub-like lower lip and a long pointed upper one
(called the galea, or beak).

x 0.8

x 2.5

85. **BUTTER-AND-EGGS**, *Linaria vulgaris*
This invasive immigrant member of the Toad Flaxes is a perennial with
creeping roots. Stems 10–30" tall bear numerous grayish-green, linear,
unstalked leaves, and spikes of spurred yellow flowers with a bearded,
pouch-like, orange lower lip. Smell is rank and unpleasant. Now widespread on roadsides, waste ground, and pastures.

86. **LITTLE MONKEY FLOWER**, *Mimulus alsinoides*
Bright, audacious monkey-faces of this dwarf Mimulus peer from rock
crevices in stony slopes from s. B.C. to n. Cal., in and w. of the Cascades.
Only 1–5" tall, the plants are conspicuous because of the brilliant yellow
flowers, with lower lip variously marked with crimson. Small leaves are
elliptical, diversely toothed, often purplish beneath. A charming wee annual.

87

x 1.7

88

x 0.6

87. LESSER PAINTBRUSH, *Castilleja attenuata*
(formerly *Orthocarpus altenuatus*)
4–8" tall, this single-stemmed hairy plant looks like a small white-topped
paintbrush, but on closer inspection this effect is seen to be due to white-
tipped bracts. The true flowers are small, yellow, and tubular, with a ring of
dark purple markings near the tube's tip. Range w. of Cascades, s. B.C. to
Cal. YELLOW LESSER PAINTBRUSH, *C. lutea* (formerly *Orthocarpus
luteus*), has larger, golden-yellow flowers and occurs e. of the Cascades.
Although the lower leaves are like those of **87**, the upper ones are broad,
with 2 small and 1 large pointed lobe.

88. TURTLE-HEAD, *Nothochelone nemorosa*
(Also known as *Chelone nemorosa*, formerly *Penstemon nemorosus*)
A deciduous perennial, with many smooth (or very sparsely-haired) stems
attaining 16–30". The thin, strongly saw-toothed, ovate leaves occur in
opposite pairs up the stems. Handsome flowers are covered with short
glandular hairs. The stamens are distinctively covered with long dense
wool. Common in open woodlands and in moist crevices of broken rock, s.
V.I. to Cal., coast to Cascades.

x 0.9

x 0.5

89. SLENDER SPEEDWELL,
Veronica filiformis
The lovely china-blue half-inch flowers
of this almost incredibly invasive plant
are becoming familiar in lawns and
boulevards, from B.C. and Wash.,
probably soon southward. Each flower
is supported by a thread-like stalk that
arises from the axil of a leaf. Leaves are
short-petaled, blunt-toothed, and
rounded. Stems creep widely.

**90. ONE-FLOWERED CANCER-
ROOT**, *Orobanche uniflora*
The solitary tube-shaped flower (some
shade of purple, very rarely yellow) is
borne at the top of an unbranched
2–4" stalk. The flower has a faint but
attractive fragrance. Its 5 lobes are
almost equal in size, and all bear 3 thin
dark stripes. Stems are leafless, for the
plant is a parasite (in the illustration,
upon the roots of **40**). Range Yukon to
Cal. and eastward.

x 0.3

91. FLY HONEYSUCKLE,
Lonicera involucrata

This upright 3–6' shrub is distinguished by 2 pairs of greenish, almost papery bracts, that become increasingly purplish, and serve as little ruffs to cup the 2 yellow flowers, or the shining black, twin berries that follow. Leaves are as much as 5" long, prominently veined, and slightly haired (especially beneath and on the veins). Range B.C. to Cal., eastward to the Atlantic, in the rich moist soil of open fields and thickets.

92. SITKA VALERIAN,
Valeriana sitchensis

This flower of moist, shaded thickets and rocky slopes is overwhelmingly scented, though the scent is not unpleasing. Small but numerous, the flowers are clustered in flat-topped arrangements, at first lavender-pink but fading white. Stems are succulent and square in section, 12–30" tall. Range Alas., to n. Cal., often montane. SCOULER'S VALERIAN, *V. scouleri*, (chiefly w. of the Cascades) 6–24", has more basal leaves with entire, not notched leaflets.

x 0.9

x 1.2

93. **WILD SCABIOUS**, *Knautia arvensis*
This immigrant from Europe is appearing
on roadsides from B.C. to Wash. and perhaps farther south. Erect but branching,
1–3' hairy stems bear pinnately lobed
6–8" leaves that are chiefly basal. 'Pin-
cushion' flower heads are various shades
of lilac and purple. The tubular corolla of
each floret is 4-lobed, and emerges from a
cupped calyx having 8–12 bristle-like
teeth.

94. **BIGROOT**, *Marah oreganus*
This high-climbing vine grows from a very
large, woody, perennial root. Large (to
8"), somewhat hairy leaves are deeply
lobed. Twisting and branched tendrils
spring from the axils of the leaves. Edible
fruit resembles small cucumbers. Range s.
B.C. to n. Cal., chiefly w. of Cascades.
BALSAM APPLE, *Echinocystis lobata*,
(annual) is similar, chiefly eastern and
southern, with more pointed leaf-lobes
and soft-prickled bladder-like fruits.

x 1.0

95

x 0.6

96

x 0.9

95. PEARLY EVERLASTING, *Anaphalis margaritacea*
The persistent, white, chaffy heads of this attractive perennial are common-ly seen on roadsides and in forest-openings all over N. Amer. The yellow florets are clustered, extremely small, and close-wrapped by parchment-like bracts. Erect, 1–2' stems and the slim alternate leaves are densely woolly. Arid-interior form has much more open, drab, buff-coloured florets.

96. SCENTLESS MAYWEED, *Matricaria maritima*
The leaf segments are more hair-like, and if the yellow disc-florets are plucked away the exposed receptacle is seen to lack chaffy scales—other-wise this immigrant plant is distinguishable with difficulty from FIELD CHAMOMILE, *Anthemis arvensis*, which has the same range and habitat, and is also an invasive weed of fields and waste places. Also similar, except for its strong odour, is STINKING CHAMOMILE, *A. cotula*.

x 0.2

x 0.25

97. NORTHWEST BALSAM-ROOT, *Balsamorhiza deltoidea*
Perennial, from a woody taproot, this plant of dry open slopes lifts each
spring a tuft of very large (to 12" long), roughly triangular leaves, and
10–30" stems. Each is topped by an ample 'flower'-head, the outer ring of
13–21 ray-florets bearing a bright yellow strap 1–2" long. Range s. V.I.,
through Puget trough to s. Cal. Replaced e. of the Cascades by silvery-
haired BALSAM-ROOT, *B. sagittata*.

98. OX-EYE DAISY, *Chrysanthemum leucanthemum*
Spreading rampantly (as is frequently observed when a species is brought to
a new country) this beautiful Eurasian plant now whitens fields and roadsides over most of N. Amer. Much smaller, numerous, white-rayed flowers
and almost yellow, pungently-aromatic, soft, pinnate leaves characterize
FEVERFEW, *C. parthenium*, a garden escapee sporadic throughout our
area.

x 0.5

x 0.5

99. **BLUE SAILORS**, *Cichorium intybus*
Another European invader of waste ground and roadsides is this coarse
1–5' plant, with its uniquely blue 'flowers.' These fold inward about noon.
The large taproot—roasted and ground—becomes chicory (mixed sometimes with coffee berries). Basal leaves are dandelion-like, and as with that
plant, cut stems exude an acrid, milky juice. Range general but chiefly w. of
the Cascades.

100. **CANADA THISTLE**, *Cirsium arvense*
'Creeping Thistle' would be a better name for this pernicious weed of
neglected fields (which spreads by wide-ranging roots) for it comes from
Eurasia. Each of the clustered, small 'flower'-heads contains florets of one
sex only—which distinguishes this from all other species of thistles (either
stamens, or pistils, are non-functional). All florets are tubular, but may be
pink, or white. Leaves fiercely spined.

x 0.2

x 0.2

101. EDIBLE THISTLE, *Cirsium edule*
The peeled stems of all thistle species are edible but these are perhaps the most mild-flavoured. The 4–6', thick, succulent stems become noticeably more slender just below the flower-heads. Deeply-toothed leaves are sparsely white-haired on both surfaces, with moderate spines along the margins and tip. From a thick, felted ball of white, cobwebby hairs just below the purple florets, project long, soft-tipped spiny scales. The pistil of each floret extends about 1/2" beyond the corolla-tube. Wet meadows and mountain slopes, Cascades west, B.C. and Wash.

102. HORSE-WEED, *Conyza canadensis*
This too-common, annual, weedy plant of waste ground all over Canada and the U.S. grows stiffly erect to as much as 4', with a great many short branchlets arising from the axils of the upper leaves. Leaves are numerous, entire, and alternate. Ray-florets are dingy-white and barely surpass the involucral bracts in length.

103

x 0.5

104

x 0.6

103. SMOOTH HAWKSBEARD, *Crepis capillaris*

This abundant (usually annual), 4–30", weed of meadows, lawns, and waste ground grows from a basal cluster of toothed, yellowish-haired but shiny leaves. Branched stems are generally smooth, with a few, smaller leaves that clasp the stem with 2 pointed 'ears.' The white 'parachute' of each seed grows directly from the top of the seed. Now found s. B.C. to Cal., chiefly w. of Cascades.

104. PINK FLEABANE, *Erigeron philadelphicus*

This widespread, 8–36", attractive species is found all over the U.S. and most of Canada, from open woods to roadside ditches and moist spots in fields. Basal leaves are variably toothed to entire; smaller, alternate stem leaves sometimes clasp the stem. Ray-florets are so slim and numerous they give the 'flower'-heads a frilly, even fragile look; they vary from deep pink to white.

x 1.0

x 0.5

105. WOOLLY SUNFLOWER, *Eriophyllum lanatum*
The bright 2" flowers of this woolly perennial appear on rocky ledges from
sea-coast to subalpine levels, B.C. to Cal., and eastward. Compact plants
are 6–12" tall, with deeply-lobed to pinnately-divided hoary leaves. Usually
there are about 12 ray-florets, but the species is highly variable.

106. GOLDENROD, *Solidago canadensis*
Over most of N. America the great plumes of Goldenrod wave in the
autumn winds. Stems, 1–6' tall, rise from creeping rhizomes, and bear
numerous, alternate, lance-linear leaves variously entire to strongly
notched. This species has many varieties, but the one illustrated is very
widespread. In the somewhat cone-shaped flower-cluster the florets appear
chiefly on the upper sides of the sub-clusters.

x 1.0

107. **TANSY**, *Tanacetum vulgare*

A 3–5', stiffly erect, dark-green, aromatic perennial, this handsome invader is now fairly common in disturbed ground, and on road margins, throughout the U.S.A. and adj. Canada. Strong stems and feather-like pinnate leaves are smooth. Many bright yellow flower-heads are arranged in a somewhat flattened cluster at the tops of the stems. Only disk-florets are present—the outermost generally pistillate, the rest perfect (with functioning stamens as well as pistil). The involucral bracts that enclose button-like flower-heads are lanceolate, thin and dry to chaffy at the tips.

x 0.8

108. **OYSTER PLANT**, *Tragopogon porrifolius*
This big purple-flowered 'dandelion' plant is cultivated in Europe for its
fleshy root, which is delicately flavoured. Now common in waste ground
and along roadsides, the plant is found over most of N. Amer. A smooth
perennial, 16–36" tall, it has erect stems clothed with slender, long-pointed
leaves. All the florets in the 'flower' are ray-florets, and all are fertile (with
both pistil and stamens). Impressive seed-heads resemble gigantic dandelion puffs. YELLOW SALSIFY, *T. dubius* occupies the same wide range
and is almost exactly similar, except that the flower-heads are lemon-yellow
instead of purple. Both species open their flowers at first light, but conceal
them about noon, when the green (sepal-like) bracts infold.

Index

Glossary

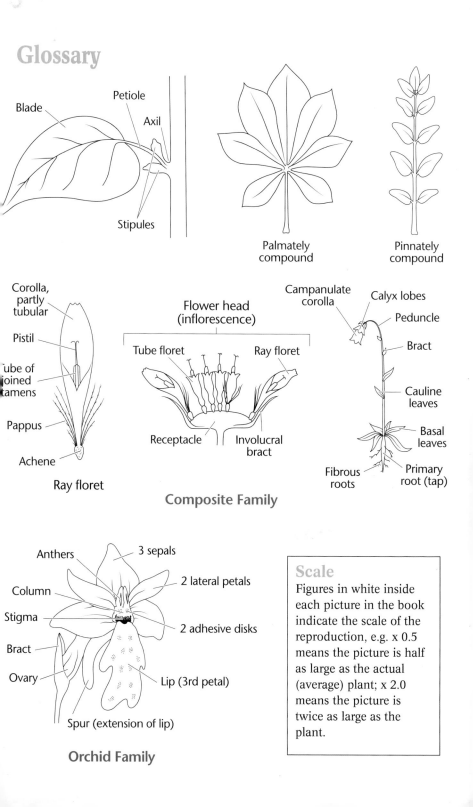

Blade

Petiole

Axil

Stipules

Palmately compound

Pinnately compound

Corolla, partly tubular

Pistil

Tube of joined stamens

Pappus

Achene

Ray floret

Flower head (inflorescence)

Tube floret

Ray floret

Receptacle

Involucral bract

Composite Family

Campanulate corolla

Calyx lobes

Peduncle

Bract

Cauline leaves

Basal leaves

Fibrous roots

Primary root (tap)

Anthers

3 sepals

Column

2 lateral petals

Stigma

Bract

2 adhesive disks

Ovary

Lip (3rd petal)

Spur (extension of lip)

Orchid Family

Scale

Figures in white inside each picture in the book indicate the scale of the reproduction, e.g. x 0.5 means the picture is half as large as the actual (average) plant; x 2.0 means the picture is twice as large as the plant.

Additional Field Guides from Harbour Publishing

Whelks to Whales: Coastal Marine Life of the Pacific Northwest by Rick M. Harbo
5.5" x 8.5" • 248 pages, 500 colour photos • 1-55017-183-6 • $24.95
This full-colour field guide to the marine life of coastal British Columbia, Alaska, Washington, Oregon and northern California is perfect for divers, boaters, beachwalkers and snorkellers.

Shells and Shellfish of the Pacific Northwest by Rick M. Harbo
5.5" x 8.5" • 272 pages, 350 colour photos • 1-55017-146-1 • $24.95
This easy-to-follow, full-colour guide introduces more than 250 species of mollusks found along the beaches and shallow waters of the Pacific Northwest.

Coastal Fishes of the Pacific Northwest by Andy Lamb and Phil Edgell
5.5" x 8.5" • 224 pages, 175 colour photos • 0-920080-75-8 • $21.95
The only comprehensive field guide to marine fishes of BC, Washington and southern Alaska.

Lake, River and Sea-Run Fishes of Canada by Frederick H. Wooding
6" x 9" • 304 pages, 50 colour illustrations and line drawings • 1-55017-175-5 • $18.95 paper
The only popular guide to freshwater fishes in all parts of Canada.

Whales of the West Coast by David E. Spalding
6" x 9" • 211 pages, 100 photos • 1-55017-199-2 • $18.95 paper
Everything you need to know about whales and dolphins of the West Coast. David Spalding brings his forty years as a naturalist to this comprehensive look at all the whales that live in Pacific Northwest waters, from the better-known orcas, grays and humpbacks to porpoises, blue whales and sperm whales.

The Beachcomber's Guide to Seashore Life in the Pacific Northwest by J. Duane Sept
5.5" x 8.5" • 240 pages, 500 colour photos • 1-55017-204-2 • $21.95
274 of the most common animals and plants found along the saltwater shores of the Pacific Northwest are described in this book. Illustrating each entry is a colour photo of the species in its natural habitat.

Pacific Seaweeds: A Guide to Common Seaweeds of the West Coast by Louis Druehl
5.5" x 8.5" • 190 pages, 80 colour photos, illustrations • 1-55017-240-9 • $24.95
The authoritative guide to over 100 common species of seaweed. Includes interesting facts, scientific information and tasty recipes.

West Coast Fossils by Rolf Ludvigsen and Graham Beard
5.5" x 8.5" • 216 pages, 250 photos, illustrations and maps • 1-55017-179-8 • $18.95
This complete new and expanded edition of a West Coast classic is a concise and thorough guide to the small and large fossils of Vancouver Island and the Gulf Islands of Canada.